NPSH for rotodynamic pumps: a reference guide

NPSH for rotodynamic pumps: a reference guide

European Association of Pump Manufacturers

ELSEVIER
ADVANCED
TECHNOLOGY

UK Elsevier Science Ltd, The Boulevard, Langford Lane, Kidlington, Oxford OX5 1GB, UK

USA Elsevier Science Inc., 665 Avenue of the Americas, New York, NY 10010, USA

JAPAN Elsevier Science Japan, Tsunashima Building Annex, 3-20-12 Yushima, Bunkyo-ku, Tokyo 113, Japan

---

First edition 1999

Library of Congress Cataloging-in-Publication Data
NPSH for rotodynamic pumps/European Association of Pump Manufacturers – 1st ed.
p. cm.
ISBN 1-85617-356-9
1. Rotary pumps – Handbooks, manuals, etc. I. European Association of Pump Manufacturers.
TJ917.N657 1999
621.6′6—dc21 99-10806
CIP

British Library Cataloguing in Publication Data
A catalogue record for this title is available from the British Library.

ISBN 1 85617 356 9

Published by
Elsevier Advanced Technology
The Boulevard, Langford Lane, Kidlington, Oxford OX5 1GB, UK
Tel: +44(0) 1865 833842
Fax: +44(0) 1865 843971

Printed in Great Britain

Contents

Preface

Over the years Europump has produced a number of books and guides to help with selecting, specifying and operating pumps and pumping systems.

This book builds on a previous Europump document on inlet conditions in pumps: *NPSH, Its Importance, Design and Test Procedures*, first published in 1971. Our new book is more comprehensive and is also more practically oriented; it is intended to be a reference guide for users of pumps and pump systems, providing a broad range of information on every aspect of cavitation and NPSH in all kinds of rotodynamic* pumps ranging from side channel to axial flow types.

This book is the result of input from many Europump-associated pump manufacturers, research and educational institutions as well as prominent European pump users. In this context we would like to express our particular gratitude to Messrs J.F. Lapray (Alstom), W. Schöffler (Sulzer Pumps), Professor B. Stoffel (Technical University of Darmstadt) and L. Teasdale (BPMA) for their valuable contribution within the Europump working group responsible for the production of this book.

For the Europump Technical Commission, 1999

Richard Canavelis — *Commission Chairman*

Pierre Fabeck — *Chairman of Working Group NPSH*

*NPSH in positive displacement pumps is not covered as it is outside the scope of this book: a companion document is due for publication later in 1999.

1. Introduction

1.1. Importance of inlet conditions

The satisfactory behaviour and performance of a rotodynamic pump frequently can be disturbed by inadequate pump inlet conditions causing cavitation. Experience has shown that insufficient pressure in the liquid at the pump inlet is a major cause of unsatisfactory performance.

The concept of the net positive suction head (NPSH) was developed for the purpose of comparing inlet conditions of the system with the inlet requirements of the pump, in order to predict suction performance. The definition of NPSH takes into account the basic parameters affecting the pump's cavitation behaviour.

1.2. Cavitation

Cavitation is the formation and sudden collapse of cavities filled primarily with vapour in a flowing liquid. The vapour cavities are formed at locations in which the pressure of the liquid drops near to the temperature-related vapour pressure. They are carried by the flow and collapse further along the flow path when the pressure again rises. These cavities may take the shape of bubbles, vortices, pockets, clouds, and so on.

The value of vapour pressure as a function of temperature (vapour pressure curve) can be obtained for many liquids from the literature (for water see Appendix A). In rotodynamic pumps, cavitation usually starts at the entry of the impeller blade, but can be found in other locations such as the leading edge of diffuser vanes, or wear rings, for example.

1.3. Effects of inadequate suction conditions

The collapse of vapour bubbles resulting from cavitation is accompanied by high levels of energy conversion, resulting in a real hammering effect on the surrounding surfaces. This can create stress fatigue in the material, plastic deformations and removal of particles from the main body. The effect can be accelerated by corrosive activity of the pumped fluid. These effects are known as 'cavitation erosion' and 'cavitation corrosion', respectively.

The extent to which cavitation can be tolerated in individual cases depends, among other things, on the pump design, the operating conditions, the characteristics of the pumped fluid, the materials from which the impeller and other parts are made, as well as the length of time that the pump is operated under cavitating conditions.

These matters are fully addressed by the manufacturer during the design of the pump and the user must give similar close attention to the NPSH available in the system at all likely operating conditions. Failure to achieve a good match of the pump and system can result in:

- noise and vibration in the pump or system pipework
- erosion and associated wear on the internal passages of the pump including impeller vanes, shafts, sleeves and even certain areas of the pump casing
- radial and axial loads which may intermittently exceed the design values of shafts, seals and bearing support systems
- a drop in the pump total head and eventually a loss of prime and subsequent cessation of pumping

- formation or accumulation of gas at the high points of the installation, including the pump itself.

2. The concept of NPSH

2.1. Pressure profile along the streamline

Figure 1 shows a typical evolution of fluid pressure in the entrance of a centrifugal pump operating at its design flow rate.

Between point 1 and A, just upstream of the inlet edge of the blade, the local absolute pressure can be estimated by using the Bernoulli equation; $p_{abs,1}$ and $p_{abs,A}$ are the absolute pressures and v_1, v_A are the absolute velocities. ΔH_{1-A} are the head losses between point 1 and point A. After point A, a further drop of pressure occurs as the fluid is accelerated and enters the blade inlet. The difference in height between point 1 and A is assumed to be zero

$$\frac{p_{abs,1}}{\rho} + \frac{v_1^2}{2} = \frac{p_{abs,A}}{\rho} + \frac{v_A^2}{2} + g\Delta H_{1-A} \tag{1}$$

The fluid specific energy at the pump inlet is expressed as

$$gH_{abs,1} = \frac{p_{abs,1}}{\rho} + \frac{v_1^2}{2} \tag{2}$$

2.2. Definition of NPSH

2.2.1. Concept of NPSH available: definition of NPSHA

Assuming that p_v is the vapour pressure at the fluid temperature, the fluid net positive suction head (NPSH) is

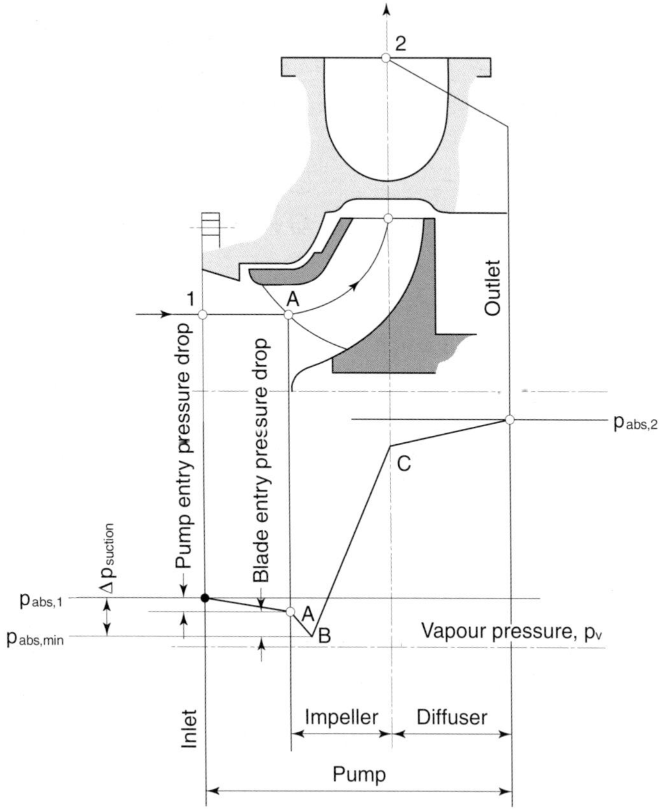

Figure 1 Pressure distribution in a pump.

defined as

$$g(\text{NPSH}) = gH_{\text{abs},1} - \frac{p_v}{\rho}$$

which represents the fluid specific energy above the vapour pressure, at the pump inlet. This can also be expressed in metres of column liquid as

$$\text{NPSH} = \frac{p_{\text{abs},1}}{\rho g} + \frac{v_1^2}{2g} - \frac{p_v}{\rho g} \tag{3}$$

The value of NPSH calculated by equation (3) is by definition called 'available' NPSH (NPSHA), as it relates to the system condition available at the pump inlet.

This quantity can be calculated at the pump inlet in all cases, since it is sufficient to know the values of $p_{abs,1}$, v_1 and p_v. If $p_{abs,1}$ and v_1 are difficult to measure at the pump inlet, this value can be calculated using the Bernoulli equation (refer to Chapter 7).

2.2.2. *Concept of NPSH required: definition of NPSHR*

Inside the pump, beacuse of the energy conversion and the internal losses, the pressure varies, as shown in Figure 1. A pressure drop between the pump inlet and the blade tip results from hydraulic losses and energy conversion, and a further pressure drop occurs after the blade tip as the fluid is accelerated again. Together, these effects make up the pressure drop between the pump inlet and the point of lowest pressure. This pressure drop is designated as $\Delta p_{suction}$

$$\Delta p_{suction} = p_{abs,1} - p_{abs,min}$$

where $p_{abs,min}$ is the value of absolute pressure corresponding to the point B where the pressure is lowest.

If the value of $p_{abs,min}$ is below the vapour pressure p_v of the liquid, vapour pockets are produced in the area where the blade pressure is below p_v: this phenomenon is commonly called **cavitation.**

For a pump impeller operating at constant flow, the way to avoid cavitation is to ensure that $p_{abs,min} > p_v$. This can be achieved by raising the value of $p_{abs,1}$.

It can be seen that $\Delta p_{suction}$ has a direct impact on the onset of cavitation: the greater $\Delta p_{suction}$ is, the more likely cavitation will occur so that a larger value of NPSHA will be needed to avoid cavitation.

The main geometrical parameters of the inlet part of the impeller governing the value of $\Delta p_{suction}$ are:

- impeller inlet eye diameter D_e (see Figure 5)
- number of blades
- thickness of blades at inlet
- inlet blade angle.

The inlet blade meridional shape and thickness profile are also major parameters that strongly influence the pressure drop $\Delta p_{suction}$.

By modification of the inlet blade design, the pressure drop $\Delta p_{suction}$ can be modified to suit the application, although there may be other effects on the overall performance of the impeller and some compromise may be necessary.

The conceptual definition of NPSH required (NPSHR) by a pump under certain conditions is the value of NPSH available (NPSHA) at which a selected phenomenon induced by cavitation starts to appear. Several criteria are used to qualify the different phenomena associated with cavitation, which lead to several definitions of NPSHR, as developed in Sections 2.3 and 2.4.

2.3. Practical definition of NPSHR

One of the most noticeable effects of cavitation is the deterioration of the pump total differential head *H* due to the choking effect of the vapour phase induced by cavitation.

As the correlation between NPSHR and the pressure drop $\Delta p_{suction}$ depends on many parameters, it is virtually impossible to calculate with accuracy the NPSHR of a pump throughout its whole operating range.

By convention, the pump industry has defined a criterion to qualify NPSHR, which is sufficient for most practical purposes. NPSHR qualified by this criterion will be referred to as NPSH3.

Basically, it corresponds to the degree of cavitation where the associated drop of pump total differential head in the first stage is equal to 3% of the 'normal' total differential head (i.e. without any cavitation, with ample suction pressure). Its value is obtained by testing. For multi-stage pumps, only the first stage total differential head is taken into consideration for determining the 3% head drop.

Figure 2 shows how the NPSH3 is determined for a given flow rate Q_a. The pump is operated under constant speed and flow rate, but with decreasing values of NPSHA, until the generated head starts to decrease. When the drop of head is equal to 3%, the corresponding value of NPSHA available is said, by definition, to be equal to the NPSH3 of the pump.

Figure 3 shows a typical NPSH3 curve for a rotodynamic pump as a function of the flow rate. At flow rates above the best efficiency point (BEP), NPSH3 usually rises steeply. Outside the operating range it is often difficult to measure NPSH3 values. This NPSH3 value is usually adopted by pump manufacturers unless otherwise required for a particular application.

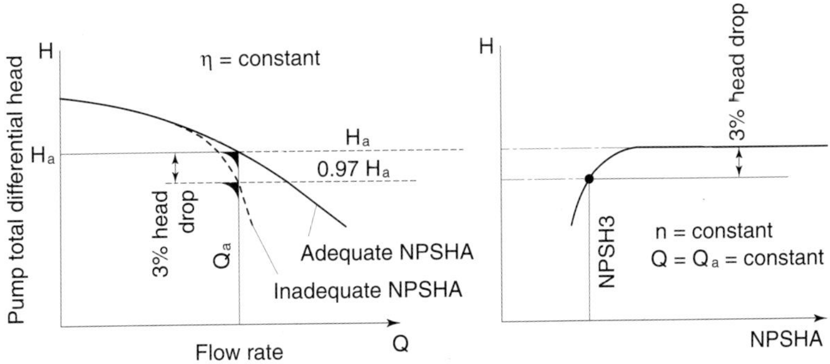

Figure 2 Practical determination of NPSHR.

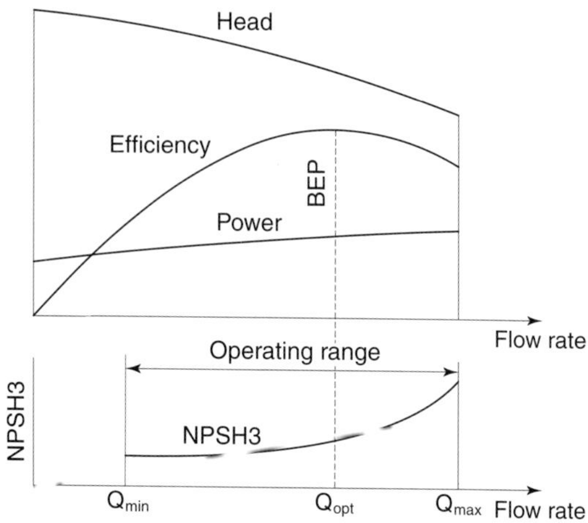

Figure 3 Typical NPSH3 curve for a centrifugal pump.

2.4. Other qualifying criteria for NPSHR

Characteristics other than total pump head can change when cavitation commences. Figure 4(a) shows, in principle, how characteristics such as the bubble length, sound level, material loss and pump total differential head can vary when the NPSHA available from the system at the pump inlet falls below that required for cavitation-free operation.

It can be seen that the peak values of the erosion tendency and the sound pressure level occur at approximately the same value of NPSHA, and that this is at an NPSHA above the value causing a fall-off of 3% in the developed head. It can also be seen that some cavitation can occur without any noticeable deterioration of the pump total head. This has led some industries to adopt more stringent criteria based on these other characteristics, when NPSH3 is not considered adequate to protect the pump during its operation. It can also be seen that some cavitation can occur without any noticeable deterioration of the pump total head. It has led to some industries adopting other criteria to qualify or define the NPSHR for a pump. Examples are:

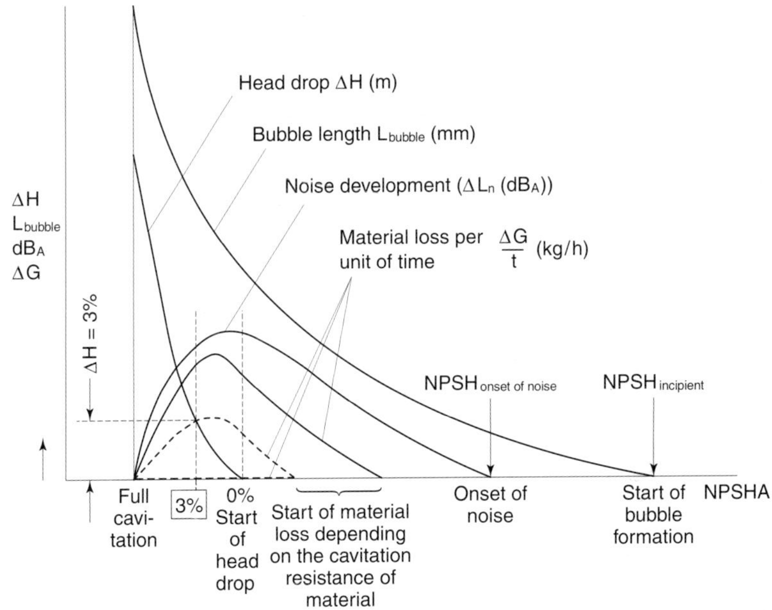

Figure 4(a) Bubble length, sound level, material loss and pump total head versus

$NPSH_{incipient}$ corresponding to incipient cavitation, i.e. to a maximum bubble length of 5 mm (also called 'cavitation-free' NPSH)

$NPSH_{onset\ of\ noise}$ corresponding to a noticeable increase in noise level

$NPSH_{material\ loss}$ corresponding to the onset of material loss

NPSH0 the lowest NPSH value at which the design pump head can be sustained

NPSH1 an NPSH value corresponding to a 1% drop of the pump head

NPSH3 an NPSH value corresponding to a 3% drop of the pump head

$NPSH_{full\ cavitation}$ at which the loss of head due to cavitation is so extensive that no change in the flow rate occurs when the resistance characteristic is reduced.

Since the sound pressure level is much easier and faster to measure than is the erosion tendency, it is frequently used to indicate avoidance of cavitation, and in high-energy pumps may be used to define an acceptable level of cavitation during normal operation. The sound emitted by a cavitating pump is very typical and is similar to what might be expected from small gravel particles passing through the pump. Visualization testing methods are often used on high-energy pumps to detect visually the onset of cavitation ($NPSH_{incipient}$). Figure 4(b) shows typical shapes of curves for $NPSH_{incipient}$, $NPSH_{onset\ of\ noise}$, $NPSH_{material\ loss}$, NPSH0, NPSH3 and $NPSH_{full\ cavitation}$.

It should be remembered that to operate with the higher NPSHR values which result from these more demanding criteria means that higher NPSHA must be provided. This can mean raising the pressure at the installation inlet or reducing the losses along the inlet pipeline, both of which introduce costs to the system. The vast majority of industrial and domestic systems are designed, and operate satisfactorily, using the NPSH3 criterion.

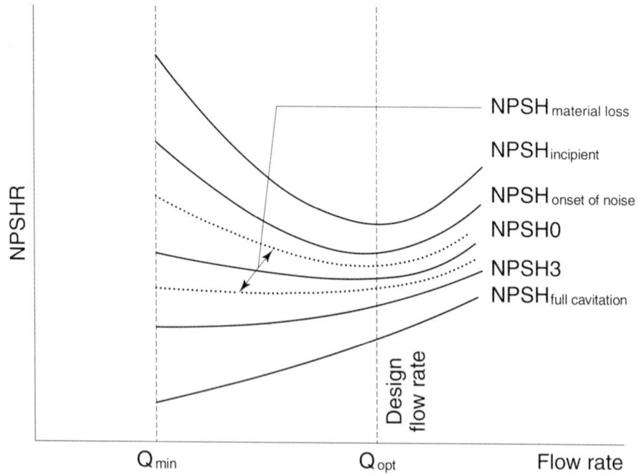

Figure 4(b) Shape of different NPSHR curves versus flow rate.

3. Governing parameters

3.1. Effect of speed and capacity

It is usually accepted that NPSHR can be expressed as follows:

$$\text{NPSHR} = k_1 \frac{c_{mA}^2}{2g} + k_2 \frac{u_A^2}{2g}$$

where c_{mA} and u_A are the meridional component of the absolute velocity and the peripheral velocity of the fluid at the inlet of the impeller blade, respectively.

For a given impeller geometry and rotational speed, u_A is fixed and c_{mA} is directly proportional to the pump flow rate. k_1 and k_2 are often assumed to be constant and depend on the geometry of the pump, as well as the operating conditions.

The above equation explains why NPSHR varies more or less as the square power of the flow rate at fixed speed, and why a change of rotational speed affects NPSHR more or less as the square of the speed ratio.

When changing rotational speed from n_1 to n_2:

$$Q_2 = Q_1 \times \left(\frac{n_2}{n_1}\right) \tag{4}$$

$$H_2 = H_1 \times \left(\frac{n_2}{n_1}\right)^2 \tag{5}$$

$$\text{NPSHR}(Q_2,\ n_2) = \text{NPSHR}(Q_1, n_1) \times \left(\frac{n_2}{n_1}\right)^2 \qquad (6)$$

Equations (4) and (5) are the classical formulae derived from the affinity laws. Equation (6) uses the recommendation from the ISO9906 standard 'Code for hydraulic performance tests – grades 1 and 2', that an exponent value of 2.0 should be used when testing for NPSH, unless otherwise specified to reflect in particular the cavitation susceptibilty of the liquid.

3.2. Effect of pump geometry

Thin, profiled vane edges at the inlet can have a beneficial effect on NPSHR as they reduce the vane 'choking' effect and thereby decrease the meridional velocity c_{mA}, the coefficients k_1 and k_2, and consequently NPSHR.

Vane inlet filing is a well-tried and simple method used in order to reduce the NPSHR requirement of a given pump.

Designs aimed at reducing NPSHR frequently result in large eye areas of the impeller, because this permits smaller values of c_{mA}. The side-effect is that large diameters of the impeller eye D_e may be detrimental to the stability of operation at partial flow rates.

Any obstruction in the suction passage upstream of the impeller vane, such as impeller locknuts and butterfly or globe valves, is detrimental to NPSHR.

The surface roughness of the impeller also plays an important role, as the smoother the surface the less disturbance to the flow and subsequently the NPSHR. This explains why, particularly on small sizes, precision-cast impellers tend to have a lower NPSHR than sand-cast ones.

3.3. Effect of impeller trim

Figure 5 illustrates the way in which NPSH3 varies when the impeller diameter is trimmed. As the pump total differential head H is reduced with the trimmed diameter, the 3% head drop applied to head H_B is reduced, it corresponds to a lesser degree of cavitation and is therefore reached earlier than with diameter D_A. Another reason is the reduction of the impeller channel length when the impeller is trimmed. Together, this explains why, at constant speed and flow, NPSH3 may increase when the impeller diameter is reduced. The extent of the increase depends on the impeller geometry. This effect is also true for NPSH0 and NPSH1.

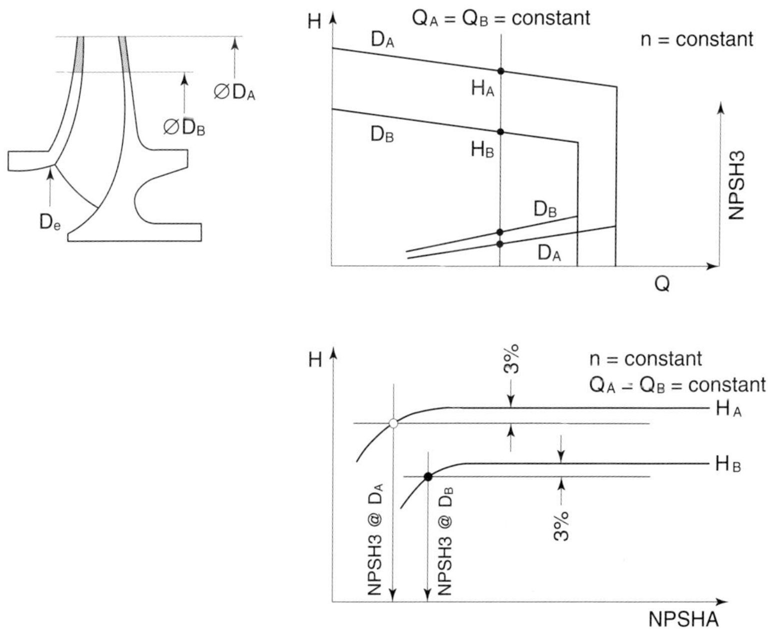

Figure 5 Effect of impeller trim.

3.4. Effect of pumped liquid

3.4.1. *Gas-to-liquid ratio*

NPSH3 values given by pump manufacturers are usually determined from tests with clear, de-aerated water, at a temperature of approximately 15°C. In the formation of cavitation bubbles, a key parameter is the gas-to-liquid ratio, that is, the ratio between the volume occupied by the vapour phase and the volume of the liquid phase. The gas-to-liquid ratio R is equal to the number of volumes of gas resulting from one volume of liquid when this volume of liquid is vaporized. For example, the value of R for water at 21°C is 53 000. For hydrocarbons, this ratio can be 10 to 100 times lower. Hot water has an R value closer to those of hydrocarbons.

The higher the gas-to-liquid ratio, the more the obstruction by the vapour cavities. This explains why cold water, with one of the highest R values of any liquid, is one of the least favourable liquids with regard to cavitation.

3.4.2. *Thermodynamic effect*

As well as the effect of the gas-to-liquid ratio, there is also a thermodynamic effect due to the specific vaporization heat of the liquid: the vaporization of the liquid tends to cool down the liquid phase, thereby lowering the vapour pressure and consequently counteracting the appearance of cavitation. Cold water is one of the least sensitive liquids to this 'beneficial' thermodynamic effect.

One can say, therefore, that the NPSHR with nearly any liquid, other than cold water, is going to be better (lower) than, or equal, to that with cold water.

Typically, NPSHR reductions due to more favourable liquids than cold water can be as much as 30%. This favourable effect is very rarely taken into account, but explains why many pumps on site do not exhibit cavitation problems although their NPSHA is lower than designed or normally required.

Appendix B, now included in the ISO9906 standard, shows how NPSHR can be reduced when handling hydrocarbon liquids or high temperature water.

3.5. Effect of inlet piping

The layout of the piping directly upstream of the pump suction flange can also influence the NPSHR of a pump. In particular, the presence or absence of anti-vortex ribs can play a role as they influence on the flow pattern at the entrance of the impeller. Also, any obstruction, restriction, elbow, valves, etc. placed upstream, but close to the pump, may affect the inlet flow pattern and subsequently the NPSHR.

In order to avoid cavitation problems, inlet pipe sizes must be at least as large as the pump inlet connection. Never use pipes or fittings on the inlet that are smaller in diameter than the pump inlet size.

Figure 6 shows recommendations for piping configurations with a minimum straight pipe length upstream of the pump. When the source comes from below the level of the reducer, horizontally mounted reducers should be eccentric and mounted with the flat side at the top in order to avoid an air trap, as is shown in Figure 6. Do not mount concentric reducers horizontally.

Bends in the suction piping of double-suction pumps should preferably be in the plane perpendicular to that of the pump shaft. If this is not possible, with the bend being in the same plane, it must be of long radius, as far as possible from the pump and due precautions taken to ensure equal, undisturbed distribution into the two eyes.

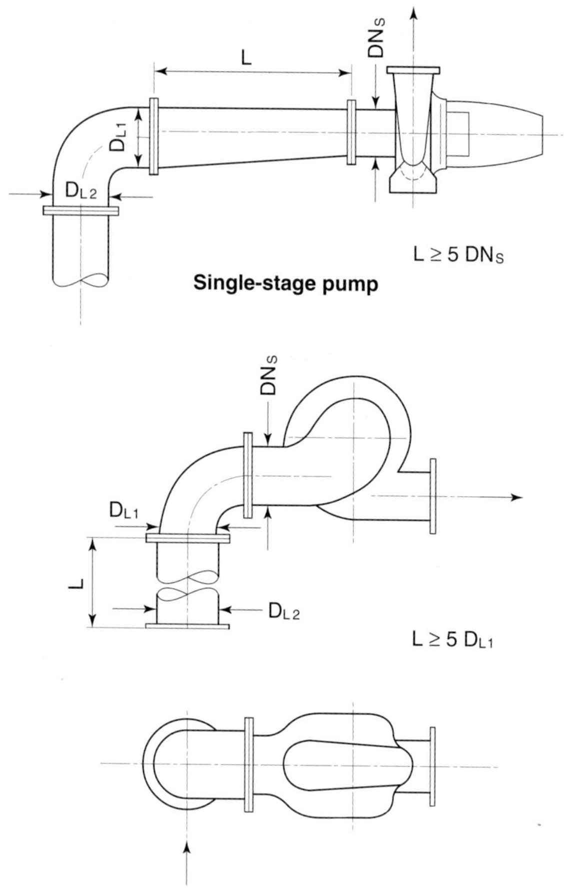

Figure 6 Recommended layout of the inlet piping.

3.6. Effect of pump material

The mechanical properties of the materials of the pump, and in particular the impeller material, play a significant role in the rate of erosion when the pump is subject to cavitation.

Under normal circumstances it is not recommended to operate a pump under cavitation. If it is unavoidable to run with cavitation, it may be beneficial to attempt to extend the impeller life by using cavitation-resistant materials.

The following properties are found to have an influence on the rate of cavitation erosion:

- the tensile strength T_S of the material
- the hardness HB of the material
- the modulus of elasticity E of the material
- the structural arrangement and constitution of the material.

Figure 7 gives an example of test results determined for various materials exposed to cavitation in a special test device. The relative resistance of the various materials shown on this illustrative graph should not be extrapolated as it also depends on the geometry of the pump, the type of cavitation and the velocity of the liquid.

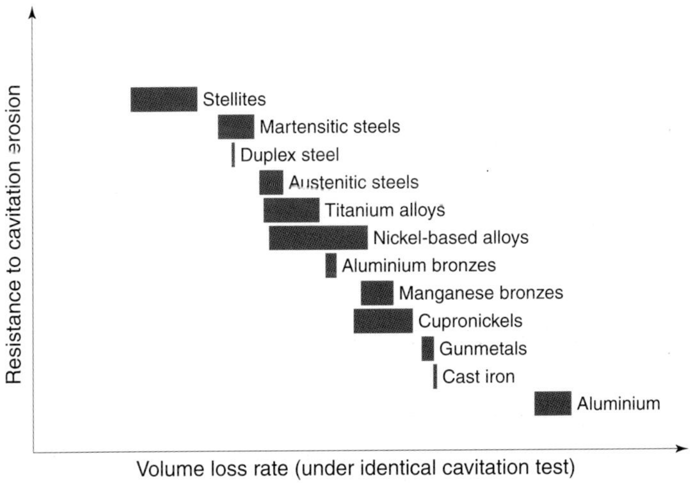

Figure 7 Effect of material characteristics.

4. Suction specific speed

It was shown in Section 3.1 (equation (6)) that the NPSHR relationship to speed was approximately similar to that of the pump total head, that is proportional to the square of the speed ratio; the concept of **suction specific speed** (n_{ss}) has been introduced to compare the suction performance of similar geometries.

By definition:

$$n_{ss} = n\frac{\sqrt{Q_{opt}}}{\mathrm{NPSH3}_{opt}^{0.75}} \tag{7}$$

where Q_{opt} is the flow rate per impeller eye at the best efficiency point, in m^3/s, and $\mathrm{NPSH3}_{opt}$ is the corresponding value of NPSH3 for the first stage at maximum impeller diameter, in metres.

This definition is similar to the specific speed n_s, defined as:

$$n_s = n\frac{\sqrt{Q_{opt}}}{H_{opt}^{0.75}} \tag{8}$$

where Q_{opt} is the flow rate per impeller eye at the best efficiency point, in m^3/s and H_{opt} is the corresponding pump total differential head per stage at maximum impeller diameter, in metres.

Note that n_{ss} and n_s depend on the SI units used for Q_{opt}, n, $\mathrm{NPSH3}_{opt}$ and H_{opt}.

The suction specific speed can also be computed in US units by using the following formula:

$$SSS = n\frac{\sqrt{Q_{\text{opt}}}}{\text{NPSH3}_{\text{opt}}^{0.75}}$$

where Q_{opt} is defined as above, but expressed in USGPM, and $\text{NPSH3}_{\text{opt}}$ is expressed in feet.

By comparing SI units to US units, one can see that: $SSS = 51.64n_{ss}$.

The hypothesis behind the suction specific speed concept is that NPSH3 follows the affinity law (i.e. proportional to the square of the speed ratio), so that, for a given pump geometry, NPSH3 can be computed at different speeds and affinity ratios by assuming that n_{SS} remains constant.

In practice, and as illustrated in Figure 8, n_{ss} improves with flow rates at constant speed, so that the hypothesis of a constant n_{ss} is not entirely correct. Typically, in SI units the n_{ss} of a centrifugal pump usually has a value between 120 and 250 (in US units approximately between 6000 and 13 000). The higher n_{ss}, the lower the NPSH3. Values of n_{ss} higher than 250 (approximately 13 000 in US units) are usually attainable only by special designs or by using inducers.

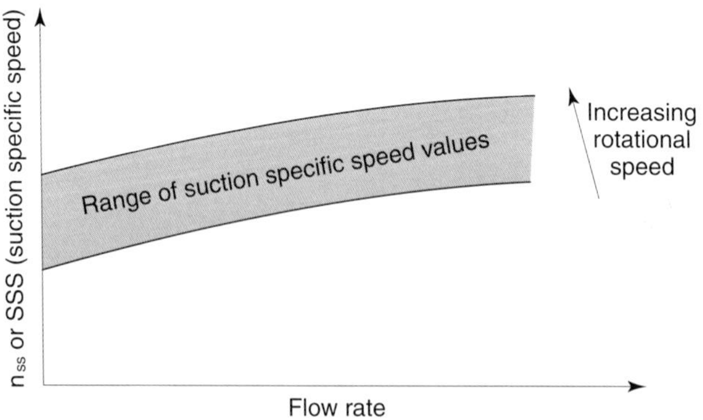

Figure 8 Suction specific speed n_{ss} (SSS) versus flow rate and speed.

When an inducer is fitted in front of the first stage impeller, satisfactory operation can be obtained with much higher values of n_{ss} (sometimes above 400, or approximately 21 000 in US units), at the sacrifice of some limitation of flow range.

Some authors have attempted to correlate the pump reliability level with the *SSS* value, suggesting that pumps with *SSS* values over 11 000 in US units (i.e. over approximately 210 in SI units) were less reliable than others. This approach, if followed blindly, is limiting, as there are several ways to optimize NPSHR without compromising the pump reliability or narrowing the operating range. It is therefore not recommended to use *SSS* or n_{ss} as a qualification parameter aimed at assessing pump reliability. Nevertheless, the suction specific speed concept is very useful in order to assess what NPSH3 value can be normally expected from a pump at a given flow rate and speed.

5. Safety margins between NPSHR and NPSHA

Equation (3) in Section 2.2 shows that one must ensure that, with the site conditions, that:

$$\text{NPSHA} > \text{NPSHR} \tag{9}$$

To obtain satisfactory operation from a rotodynamic pump, a minimum NPSHA must always be secured in order to avoid or restrict cavitation to an acceptable level. The pump manufacturer may recommend a safety margin to be applied between the pump NPSH3 and the NPSHA.

The principal parameters governing the safety margin are:

- pump size, type and specific speed
- the power concentration of the pump, related essentially to the circumferential velocity at the impeller entry
- impeller design, and in particular the meridional shape and the geometry of the blades
- the pumped liquid and the temperature, and whether a boiling point or range is encountered with the fluid
- the resistance of the material to cavitation erosion.

One rule of thumb followed by several standards, and by the pump industry generally, is that the site conditions should be at least such that

$$0.5\ \text{m} + \text{NPSH3} \leq \text{NPSHA} \tag{10}$$

that is, there should be at least a 0.5 m margin between NPSH3 and NPSHA. This rule can lead to wrong selection in some applications, as the value (NPSH3 + 0.5 m) may coincide with the zone of maximum erosion rate, as shown in Figure 9.

Some industries have, therefore, elected to use a safety coefficient (multiplier) S_A to determine the recommended margin, so that the minimum NPSHA required can be calculated as follows:

$$\text{NPSHA} \geq \text{NPSH3} \times S_A \tag{11}$$

The safety coefficient S_A must always be given by the pump manufacturer for each application, based on the manufacturer's experience. Figure 10 illustrates how the safety coefficient S_A varies versus NPSH3 at the best efficiency point.

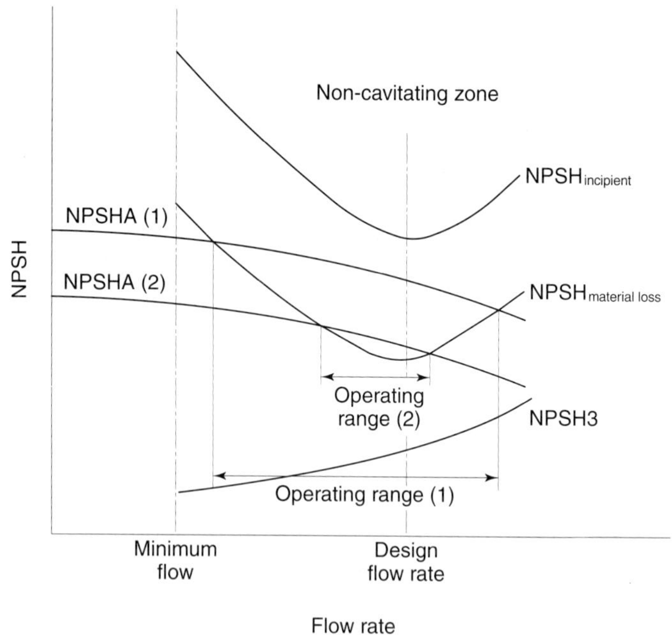

Figure 9 Margins between NPSHR and NPSHA.

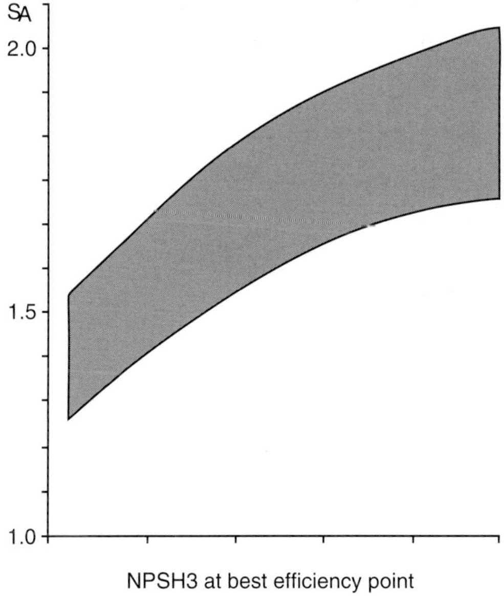

Figure 10 Safety coefficient S_A versus NPSH3.

When the pump is operating at partial flow, a higher coefficient is desired, and the manufacturer must be consulted for the selection of the appropriate safety coefficient.

The pump material must also be taken into account, as some materials exhibit a higher resistance to cavitation than others (refer to Section 3.6).

It is usually accepted though that most pump applications can tolerate some small level of cavitation without any adverse consequences. For high-energy pumps, such as boiler-feed pumps, it is sometimes necessary to use the very penalizing criterion: $NPSHA > NPSH_{incipient}$, that is to impose a totally 'cavitation-free' operation.

Another aspect which is a frequent cause of cavitation problems on pump installations is the operation at a capacity higher than designed. As shown in Figure 9, NPHSA always decreases with higher flow rates, while NPSHR increases so that even a modest increase in the flow rate may result in the onset of cavitation.

This situation will occur typically whenever the actual head losses of the system are smaller than originally estimated. It is therefore recommended to always check that the pump is operating close to the design condition. A simple way to achieve this is by comparing the power readings on-site with the pump curve.

The slope of the NPSH curves also play an important role in determining how an increase in flow rate can affect the NPSH margin. As seen in Figure 11, for two cases (solid lines and dotted lines), where the NPSH margin is identical at design flow rate *Q*, the incremental Δ*Q* needed to annihilate this NPSH margin is very dependent on the slope of the NPSHR and the NPSHA curves.

Figure 9 also illustrates how two different situations of NPSHA, referenced by NPSHA(1) and NPSHA(2), can impact on the operating range of the pump when the criterion of no material loss is selected.

Some manufacturers recommend increasing the NPSH3 readings derived from their published curve by some value (usually

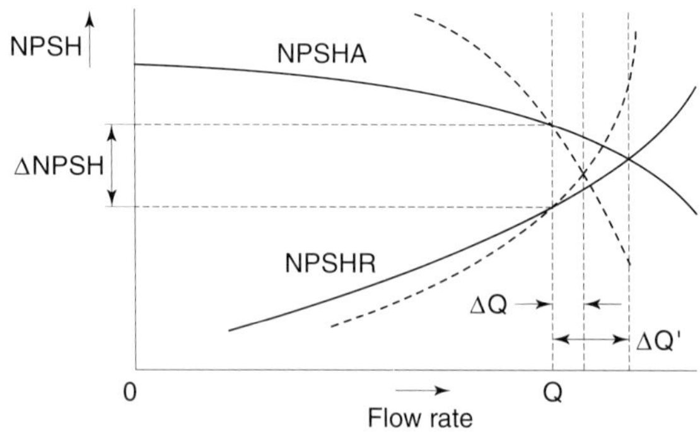

Figure 11 Effect of NPSH's curve shape.

0.5 m), when a guaranteed NPSH3 value is requested. It should be clear that this should not be confused with the 0.5 m margin frequently recommended between NPSH3 and NPSHA.

Also, it has been shown in Section 3.4 that the cavitation characteristics of clean, cold water are more demanding than those of other commonly pumped fluids. Thus, a system designed on the basis of the NPSH3 derived from tests on clean, cold water but pumping another fluid will usually give a higher safety margin than indicated. As was stated in Section 3.4, most users do not make adjustment to take account of this additional margin, although it plays a favourable and frequently significant role in most applications. There are also some liquids that will require a higher NPSH margin than does clean, cold water, and this will have the effect of reducing the indicated safety margin. Any correction to be applied as a result should be considered by the pump manufacturer and user together.

6. Practical determination of NPSHR

Several standards can be applied to the testing of NPSHR for centrifugal pumps. A good example is the ISO standard ISO9906 'Rotodynamic pumps – code for hydraulic performance tests for acceptance – grades 1 and 2' published in 1996.

Testing to grade 2 gives more tolerance than to grade 1. Appendix C is an abstract from ISO9906 dealing with NPSH testing.

Performing a NPSH test is, in fact, equivalent to measuring a value of NPSHA corresponding to the 3% head drop criterion

$$\mathrm{NPSHA} = H_1 + \frac{p_{\mathrm{amb}} - p_v}{\rho g}$$

The atmospheric pressure p_{amb} and vapour pressure p_v are relatively easy to obtain – provided the precautions stated in Sections 7.1 and 7.3 are considered – and the main point is to measure the total relative suction head H_1.

Different scenarios arise during testing, depending on the combination of two factors:

- whether pressure measurements are taken from Bourdon gauges, from column gauges or from pressure transducers
- whether the suction line is under pressure or under vacuum.

Also, in the case of column gauges, the density of the liquid in the column must be taken into account.

Appendix D details how H_1 should be measured in the various possible testing configurations.

Some industries or customers may refer to their own methods for NPSH testing, but the relationship or guidelines given in this document remain applicable. Control of inlet pressure during testing may be achieved either by throttling the inlet line or by applying a vacuum to the inlet. The latter method is preferable, as it causes less flow disturbance and tends to reduce the amount of dissolved air or other gases. However, this method is not always easy to apply and the throttling method can give adequate, reliable results.

7. Determination of NPSHA

In the NPSHA equation, if we use the NPSH plane as the reference plane, we find that

$$\text{NPSHA} = H_1 + \frac{p_{\text{amb}} - p_v}{\rho g}$$

where H_1 is the total relative head at the pump inlet, defined by

$$H_1 = z_1 + \frac{p_1}{\rho g} + \frac{{v_1}^2}{2g}$$

According to the Bernoulli equation, it is possible to determine this total relative head at the pump inlet by measuring the total relative head at another point in the system, upstream from the pump, and by deducting the total head losses J_1 between this point and the pump inlet (see Figure 12).

According to Bernoulli's equation $H_1 = H_0 - \Delta H_{0-1}$.

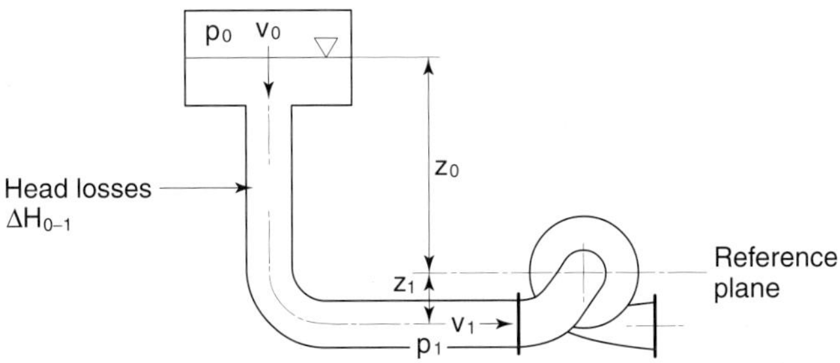

Figure 12 Calculation of NPSHA.

It is convenient to chose point '0' such that $V_0 \simeq 0$ so that:

$$\text{NPSHA} = z_0 + \frac{p_0 + p_{\text{amb}} - p_v}{\rho g} - \Delta H_{0-1} \qquad (12)$$

and this method eliminates the need for computing the velocity head $v_1^2/2g$, but creates the need for determining the head losses ΔH_{0-1}.

The determination of NPSHA is usually made by those responsible for the system design such as end users or engineering companies. It must be carefully calculated and communicated to the pump manufacturer in order to select the most suitable pump.

In the following sections we list a few points that are often subject to misinterpretation.

7.1. Atmospheric pressure

It is usual for pressure in a pump and system to be expressed as gauge pressure: values for 'head' are gauge value, that is above or below atmospheric pressure. By contrast, the vapour pressure of a liquid is always expressed as absolute pressure.

Since NPSH is the margin or difference between the actual static pressure and the vapour pressure, these must be brought to the same form if an incorrect – usually unnecessarily low – value of NPSHA is to be avoided.

The correction is best made by adding atmospheric pressure to the static pressure gauge reading and then deducting the vapour pressure.

All heights must be expressed by using the correct density of the liquid, and the atmospheric pressure must be corrected for the altitude of the site.

7.2. Reference level

By definition the NPSHR is based on the elevation referred to the NPSH reference plane (see below). As NPSHA is usually referred to the level of the inlet connection, which may be different from that of the NPSH reference plane, an eventual correction of the level must be applied when comparing NPSHA and NPSHR, or when correcting for rotational speed.

The NPSH reference plane is the horizontal plane through the centre of the circle described by the external points of the entrance edges of the impeller blades; in the case of double inlet pumps with vertical or inclined axis, it is the plane through the higher centre. The manufacturer should indicate the position of this plane with respect to precise physical reference features on the pump.

Figure 13 shows how the reference plane is defined in various configurations. For vertical double-suction impellers, the NPSH must be referred to the entrance of the upper suction eye.

In most cases, process engineers do not know at the time of calculating the NPSHA what size of pump will be used (this determines the height of the pump inlet centreline versus datum), so that the NPSHA is usually referenced to the floor datum. When selecting a pump size, a correction on the NPSHA is needed in order to convert it from the floor level to the pump centreline. Imprecision or confusion about the reference level is a frequent cause of misunderstanding and incorrect pump selection.

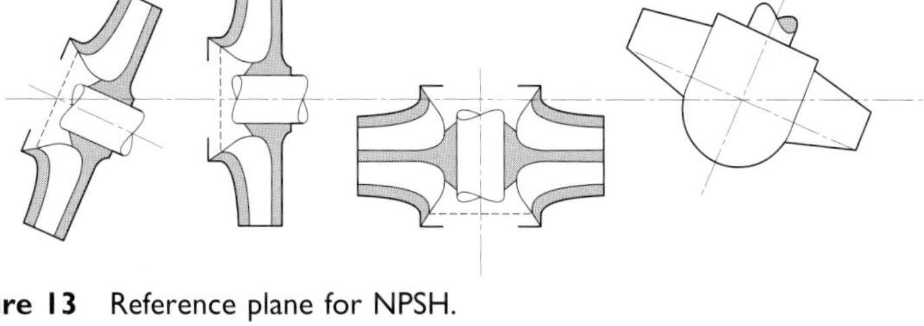

Figure 13 Reference plane for NPSH.

7.3. Temperature and vapour pressure

Both the NPSHA and the vapour pressure must be calculated at the temperature of the liquid at pump entrance.

Any imprecision in the value of vapour pressure/temperature will result in a wrong value of NPSHA and may result in inappropriate pump selection. The liquid cools down in most processes and in many cases a temperature higher than actual is specified. Inversely, long suction lines may be exposed to heating which could affect the vapour pressure at the pump inlet.

7.4. Velocity head

The NPSHA must include the velocity head component $v^2/2g$ at the pump suction. Typically, $v = 2–5$ m/s at the entrance of a pump which can result in differences of:

- $v^2/2g = 0.2$ m, at 2 m/s
- $v^2/2g = 1.25$ m, at 5 m/s.

If NPSHA is determined from measurement on-site, the Bourdon gauges will give the value of **static** inlet head, to which the dynamic velocity head $v^2/2g$ must be added. If NPSHA is calculated from the pressure in a pressure vessel or tank upstream of the pump, and where fluid velocity is close to zero (such as Figure 11), then the velocity head at this point does not need to be computed.

7.5. Measurements from Bourdon gauges

If the static inlet pressure is measured from Bourdon gauges, two points must be checked:

- if the suction pipe is under pressure, the level difference between the gauge and the reference plane must be corrected for, as the gauge line is full of liquid
- if the suction pipe is under vacuum, the level difference between gauge and reference plane must **not** be corrected for, as the gauge line is full of air. Careful checks should be made in this case to ensure that no leaks are present.

7.6. Conversion from pressure to head

All values of line pressure, atmospheric pressure and vapour pressure must be converted into an equivalent height of liquid column using the following formula:

$$H = \frac{p}{\rho g}$$

where ρ is the density in kg/m^2, g is the gravity constant (usually equal to 9.81 m/s^2), H is in metres of LC (liquid column) and p is the pressure in pascals. It is essential to convert by using the fluid's correct value of density, which varies with temperature.

An easy method for conversion is to use the relative density of the liquid versus water and apply it to the following: 1 bar is equivalent to 10.2 m of H_2O at 15°C.

Appendix E can also be used when pressure values must be converted into different units.

8. Examples of calculation of NPSHA

The following examples are representative for most situations likely to occur.

8.1. Pump under positive head and liquid under atmospheric pressure

$$\mathrm{NPSHA} = z_0 + \frac{p_0 + p_{\mathrm{amb}} - p_v}{\rho g} - \Delta H_{0-1}$$

An example (see Figure 14): $z_0 = +3$ m (pump under load, hence z_0 positive), water at 80°C when $p_v = 0.474$ bar or 47 400 Pa, $\rho = 973$ kg/m^3.

Atmospheric pressure

$$p_{\mathrm{amb}} = 750\,\mathrm{mmHg} \times 133.322 = 100\,000\,\mathrm{Pa}.$$

The head loss at the outlet from the pond and in the conduit is

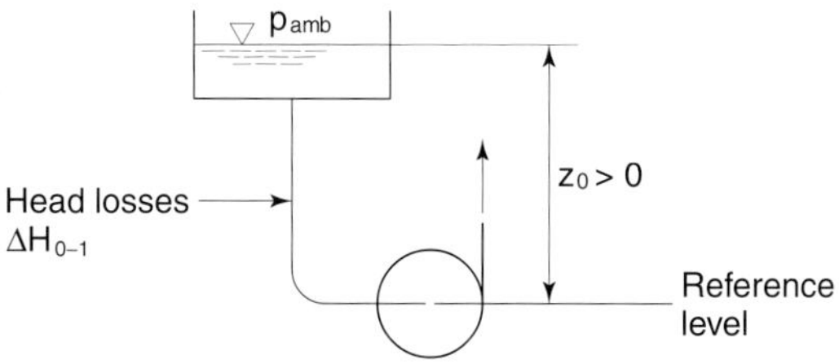

Figure 14 Pump under positive head and liquid under atmospheric pressure.

1.5 m. $p_0 = 0$ since there is no excess pressure on the free liquid level

$$\text{NPSHA} = +3\ \text{m} + \frac{100\ 000 - 47\ 400}{973 \times 9.8066}\ \text{m} - 1.5\ \text{m} = 7\ \text{m}$$

8.2. Pump under negative head and liquid under atmospheric pressure

$$\text{NPSHA} = z_0 + \frac{p_0 + p_{\text{amb}} - p_v}{\rho g} - \Delta H_{0-1}$$

An example (see Figure 15): $z_0 = -4$ m (pump under suction, hence z_0 is negative), water at 20°C, hence $p_v = 0.023$ bar or 2340 Pa, $\rho = 999\ \text{kg/m}^3$. Atmospheric pressure: $p_{\text{amb}} = 1020\ \text{mbar} \times 100 = 102\ 000$ Pa.

Head losses in the filter disk and in the pipeline: $\Delta H_{0-1} = 1.2$ m; $p_0 = 0$, since there is no excess pressure on the free liquid level.

$$\text{NPSHA} = -4\ \text{m} + \frac{102\ 000 - 2340}{999 \times 9.8066}\ \text{m} - 1.2\ \text{m} = 5\ \text{m}$$

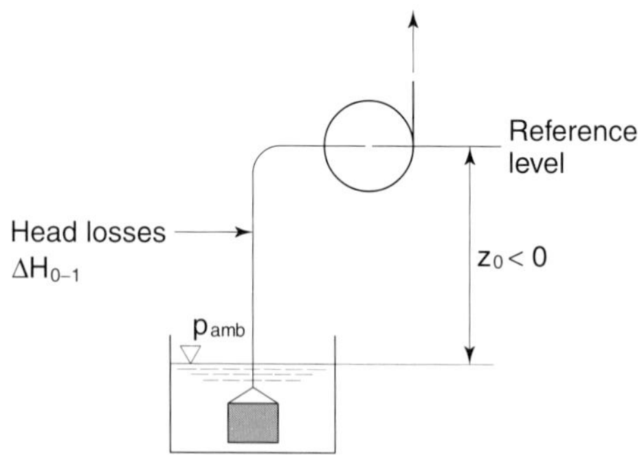

Figure 15 Pump under negative head and liquid under atmospheric pressure.

8.3. Liquid level under positive pressure and pump below the liquid level

It should be noted that a pressure gauge always indicates a gauge pressure, that is to say, with reference to the atmosphere:

$$\text{NPSHA} = z_0 + \frac{p_0 + p_{\text{amb}} - p_v}{\rho g} - \Delta H_{0-1}$$

An example (see Figure 16): $z_0 = +3\,\text{m}$ (water level above the reference plane), water at 80°C. Hence $p_v = 47\,400\,\text{Pa}$ (see Section 8.1), $\rho = 973\,\text{kg/m}^2$. Atmospheric pressure

$$p_{\text{amb}} = 750\,\text{mmHg} \times 133\,322 = 100\,000\ \text{Pa}.$$

p_0 = effective pressure indicated on the pressure gauge = 2 kgf/cm^2. This pressure is obtained as the result of gas being present in the intake tank.

$$p_0 = 2\,\text{kgf/cm}^2 \times 98\,066 = 196\,000\ Pa.$$

Head loss at the outlet from the tank and in the pipeline $\Delta H_{0-1} = 2$ m.

$$\text{NPSHA} = +3\ \text{m} + \frac{196\ 000 + 100\ 000 - 47\ 400}{973 \times 9.8066}\ \text{m} - 2\ \text{m}$$
$$= 27.1\ \text{m}$$

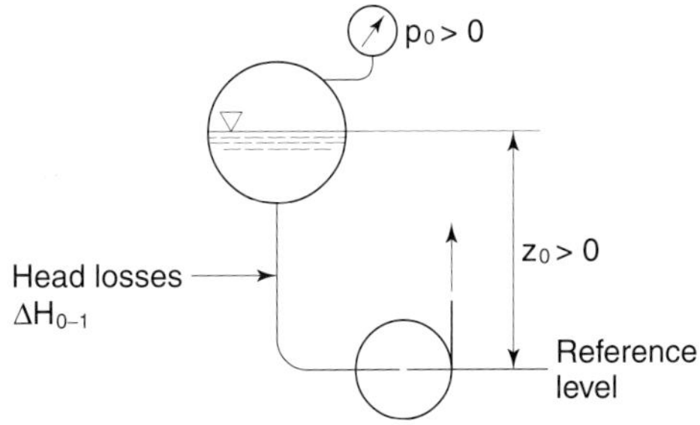

Figure 16 Liquid level under positive pressure and pump below the liquid level.

8.4. Liquid level under positive pressure and pump above the liquid level

$$\text{NPSHA} = z_0 + \frac{p_0 + p_{\text{amb}} - p_v}{\rho g} - \Delta H_{0-1}$$

We will take the same example as above (see Figure 17), but noting that z_0 is negative.

Assume $z_0 = -2$ m. As above, we obtain: water at 80°C, when $p_v = 47\,400$ Pa. $\rho = 923$ kg/m^3, $p_{\text{amb}} = 100\,000$ Pa and $p_0 = 196\,000$ Pa. Head losses in the pipeline, including the inlet: $\Delta H_{0-1} = 2$ m

$$\begin{aligned}\text{NPSHA} &= -2\text{ m} + \frac{196\ 000 + 100\ 000 - 47\ 400}{973 \times 9.8066}\text{ m} - 2\text{ m} \\ &= 22.1\text{ m}\end{aligned}$$

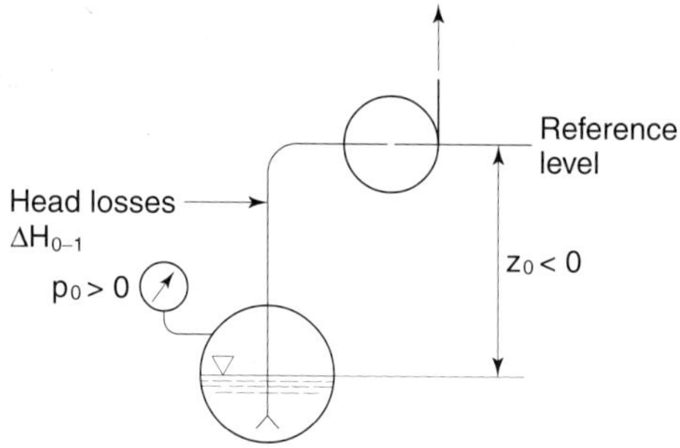

Figure 17 Liquid level under positive pressure and pump above the liquid level.

8.5. Liquid level under negative pressure and pump below the liquid level

It is understood that the vacuum must not in any event allow boiling of the liquid. The vacuum gauge usually indicates the vacuum with reference to atmospheric pressure, and must take a minus sign, unless one uses a special type of gauge measuring the absolute pressure. The value read off must take the minus sign

$$\mathrm{NPSHA} = z_0 + \frac{p_0 + p_{\mathrm{amb}} - p_v}{\rho g} - \Delta H_{0-1}$$

An example (see Figure 18): $z_0 = +3$ m (liquid level above the reference plane). Water at 80°C, hence $p_v = 47\,400$ Pa. $\rho = 973$ kg/m^3. Atmospheric pressure:

$$p_{\mathrm{amb}} = 750\,\mathrm{mmHg} \times 133\,322 = 100\,000\ \mathrm{Pa}.$$

$$p_0 = -0.3\ \mathrm{kgf/cm^2} \times 98\,066$$
$$= -29\,720\ \mathrm{Pa}\ \text{(negative value)}.$$

$z_0 = +3$ m. $\Delta H_{0-1} = 1$ m = head loss at the outlet from the tank and in the pipeline.

$$\mathrm{NPSHA} = +3\ \mathrm{m} + \frac{-29\,420 + 100\,000 - 47\,400}{973 \times 9.8066}\ \mathrm{m} - 1\ \mathrm{m}$$
$$= 4.43\ \mathrm{m}$$

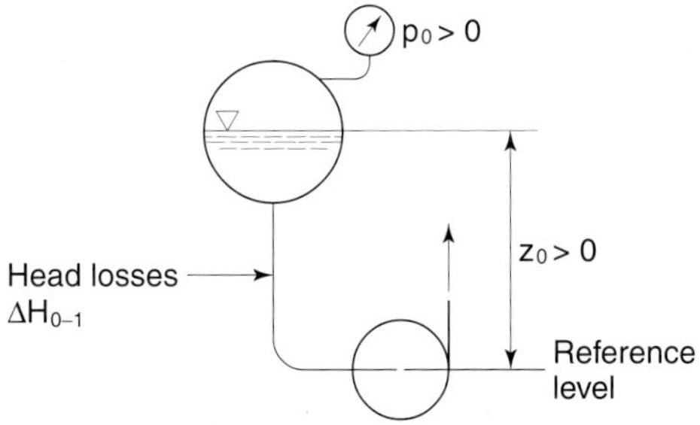

Figure 18 Liquid level under negative pressure and pump below the liquid level.

8.6. Liquid level under negative pressure and pump above the liquid level

As above, it is understood that the vacuum must permit boiling of the liquid.

The vacuum gauge always indicates vacuum with reference to atmosphere. The value read off must take a minus sign.

$$\text{NPSHA} = z_0 + \frac{p_0 + p_{\text{amb}} - p_v}{\rho g} - \Delta H_{0-1}$$

An example (see Figure 19): $z_0 = -3$ m (negative value), water at 20°C, hence $p_v = 2340$ Pa, $\rho = 999$ kg/m^3. Atmospheric pressure:

$$p_{\text{amb}} = 1020 \text{ millibars} \times 100 = 102\,000 \text{ Pa}$$

$$\begin{aligned} p_0 &= \text{reading from the vacuum gauge} \\ &= -0.2\text{kgf/cm}^2 \times 98\,066 \\ &= -19\,600 \text{ Pa}, \\ \text{or} &= -0.196 \text{ bar} \times 100\,000 \\ &= -19\,600 \text{ Pa} \end{aligned}$$

$z_0 = -2$ m (negative value), $\Delta H_{0-1} = 1$ m = head loss at the inlet of the pipe and in the pipe itself.

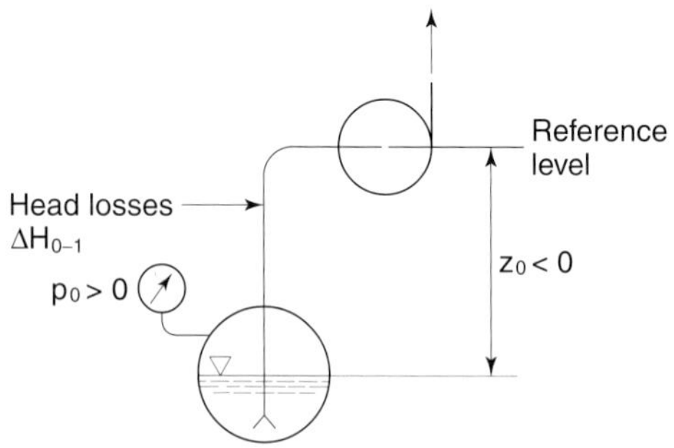

Figure 19 Liquid level under negative pressure and pump above the liquid level.

$$\text{NPSHA} = -2\text{ m} + \frac{-19\,600 + 102\,000 - 2340}{999 \times 9.8066}\text{ m} - 1\text{ m}$$
$$= 5.2\text{ m}$$

8.7. Liquid level under the vapour pressure of the liquid

The formula is simplified, and we obtain:

$$\text{NPSHA} = z_0 - \Delta H_{0-1}$$

Note that the NPSHA is independent of the vapour pressure of the pumped liquid and its temperature, assuming that there is no temperature difference in the liquid between the drum and the pump.

Attention must be paid, however, to temperature variations as a function of time and if these variations are rapid, the problem must be put to experts.

An example (see Figure 20): water at 80°C, subjected to its vapour pressure, that is an absolute pressure of 0.473 bar. $z_0 = +3$ m (always positive), $\Delta H_{0-1} = 1$ m – head loss at the outlet from the tank and in the pipe:

$$\text{NPSHA} = 3\text{ m} - 1\text{ m} = 2\text{m}$$

Note should be taken of the case where the head loss ΔH_{0-1} is negligible. In this case, the available NPSHA is equal to the geometric height of the head.

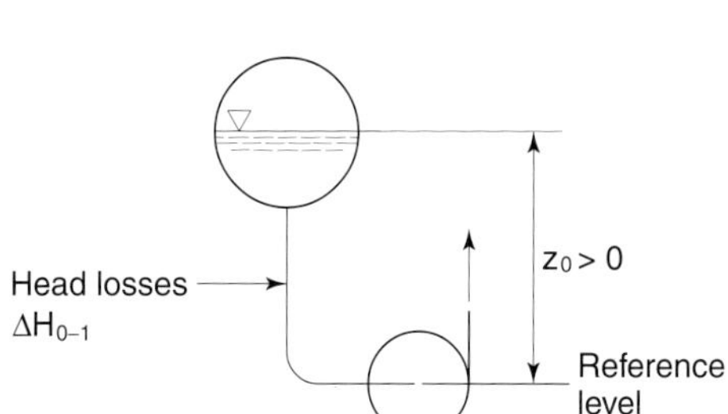

Figure 20 Liquid level under the vapour pressure of the liquid.

9. Other phenomena

9.1. Suction recirculation

When a centrifugal pump operates at low flow rates, there is a mismatch between the inlet flow pattern and the blade angles so that internal recirculation may take place. Figure 21 shows a typical suction flow pattern when suction recirculation occurs.

Two different types of internal recirculation can take place: suction and/or discharge recirculation. Both types of recirculation are associated with increased noise and vibration levels and with danger of erosion damage similar, at first sight, to that caused by cavitation.

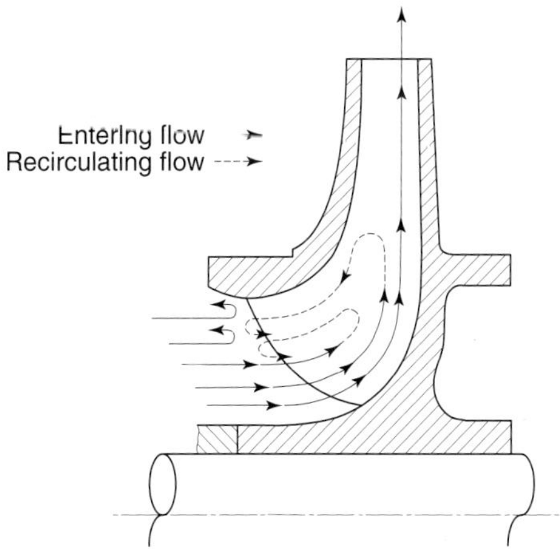

Figure 21 Suction recirculation.

Cavitation erosion damage caused by suction recirculation usually occurs at the inlet tip of impeller blades. Both suction and pressure sides of the blades can be affected, depending on impeller geometry and layout of upstream piping. It is sometimes difficult to distinguish this type of damage with that resulting from pressure-side cavitation induced by operation above the design flow rate. Suction recirculation can also cause damage on structures upstream of the impeller. The cavitation induced by suction recirculation can be suppressed by an increase of suction pressure, but it will take a higher NPSH margin to do so than that required to suppress classical cavitation.

9.2. Corrosion

Corrosion can take various forms according to the kind of corrosive attack and is usually visible on all parts in contact with the liquid.

It has been noted that cavitation can accelerate the damage caused by corrosion so that the phenomena can amplify each other and accelerate the destruction of the impeller.

9.3. Submergence under a free surface

The submergence, *S*, is defined as the difference in level between the free surface of the liquid and the entrance in the suction bell or the suction line.

When pumping liquid from a reservoir under atmospheric pressure, there is a possibility of formation of a vortex that will entrain air by friction. This happens when a high acceleration of the liquid occurs close enough to the liquid level, that is when the submergence *S* is insufficient to prevent vortex formation.

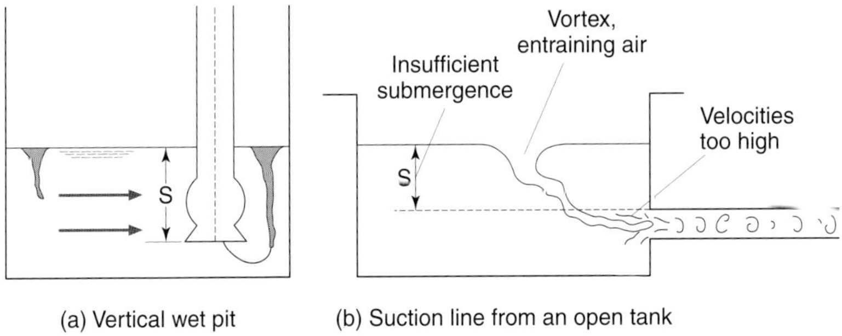

Figure 22 Minimum required submergence.

Several remedies, consisting, for example, of 'vortex breakers', are possible to avoid formation of vortices. The principle is always the same: create a physical obstacle along the vortex so that it cannot develop.

It must be noted that this phenomenon can occur with vertical wet-pit pumps (Figure 22a) but also in a suction line drawing liquid from a reservoir (Figure 22b).

9.4. Air or gas entrainment

If a high volumetric content of air or gas is present in the pumped liquid, this may cause a deterioration of the pump performance that must not be confused with cavitation.

It is generally agreed that, starting from ~2% of gas content by volume, some deterioration of the pump performance starts to appear and that the maximum possible air or gas content that a conventional centrifugal pump can handle is typically between 5% and 10%, beyond which it will progressively lose prime and stop pumping. Special designs such as inducers can be produced to tolerate much higher gas content.

9.5. Dissolved gases

Most fluids will take up gases into solution. This may be air or any other gas with which fluid has at some time interfaced.

The ability of a liquid to retain dissolved gases depends on the substances concerned, the temperature of the solution and the environmental pressure. Raising the temperature and/or lowering the pressure can cause dissolved gas to be driven off. Since these are the conditions which can cause vapour bubbles to be formed, there is often confusion between the two effects, and NPSH testing is more accurate if carried out with degasified water. Experience has shown that an air content in water corresponding to 5 ppm of dissolved oxygen or less leads to reliable and repetitive NPSHR measurements.

The essential difference is that restitution of the pressure and temperature regime does not cause the immediate re-solution of the gas bubble. Dissolving the gas requires a finite time and the bubble is reduced gradually. This compares with the instantaneous reduction in volume resulting from condensation of the liquid vapour. Thus gas bubbles reduce the implosion energy of cavitating flows, although they cause a similar drop in performance through their 'blocking' effect.

In practice, the formation of vapour due to cavitation will usually be accompanied by the generation of gas, forming in the same bubble. The gas content will reduce the destructive impact of the cavitation but only partially. Where low energy cavitation has been experienced, its effects have been mitigated to some degree by deliberately introducing a stream of air bubbles into the affected zone.

10. Conclusions

This book was developed by Europump to present the concept of NPSH together with methodologies for its application. The overall purpose was to bring about greater understanding of the various phenomena associated with NPSH. It has been written with the intention of reducing the risks of mismatch between the pump and the installation in which the pump must operate. Most graphs and figures intentionally have no scale as the aim is to illustrate the concepts rather than to be a design reference.

11. Terms, symbols and units

Table 11.1 In accordance with ISO9906 and EN12723

Symbol	Definition	Dimension	Unit
c_{mA}	meridional component of the absolute velocity at the impeller inlet blade (see Figure 1)	LT^{-1}	m/s
g	Acceleration due to gravity (usually equal to 9.81 m/s^2, but can change with site conditions)	LT^{-2}	m/s^2
H	Pump total differential head: $H = H_2 - H_1$	L	m
H_1	Total relative head in the inlet section of the pump $$H_1 = z_1 + \frac{p_1}{\rho g} + \frac{v_1{}^2}{2g}$$	L	m
H_2	Total relative head in the outlet section of the pump $$H_2 = z_2 + \frac{p_2}{\rho g} + \frac{v_2^2}{2g}$$	L	m
$H_{abs,1}$	Total absolute head in the inlet section of the pump	L	m
n	Rotational speed	T^{-1}	min^{-1}

Symbol	Definition	Dimension	Unit
n_s	Specific speed of the pump, in SI units $$n_s = n\frac{\sqrt{Q_{opt}}}{H_{opt}^{0.75}}$$ where Q_{opt} is the flow rate per impeller eye at best efficiency point, in m^3/s and H_{opt} is the corresponding pump total differential head per stage at maximum impeller diameter, in metres	$L^{0.75}T^{-1.5}$	–*
NPSH	Net positive suction head	L	m
NPSHA	NPSH available at the pump inlet	L	m
NPSHR	NPSH required at the pump inlet	L	m
NPSH3	NPSHR qualified by the 3% head drop criterion to the first stage impeller	L	m
$NPSH_{incipient}$†	NPSHR corresponding to the onset of cavitation (i.e. cavitation-free or incipient condition)	L	m
$NPSH_{onset\ of\ noise}$†	NPSHR corresponding to a noticeable increase in noise level	L	m
$NPSH_{material\ loss}$†	NPSHR corresponding to the onset of material loss	L	m
NPSH0†	NPSHR corresponding to the lowest NPSHA value at which the pump total head can be sustained (i.e. corresponding to a 0% head drop)	L	m
NPSH1†	NPSHR qualified by the 1% head drop criterion to the first stage impeller	L	m
$NPSH_{full\ cavitation}$†	NPSHR at which the loss of head due to cavitation is so extensive that no change in the flow rate occurs when the hydraulic resistance is reduced	L	m

*The numerical value of this parameter depends of the choice of the units.
†This definition may be used in certain applications.

Symbol	Definition	Dimension	Unit
n_{ss}	Suction specific speed of the pump, in SI units $$n_{ss} = n\frac{\sqrt{Q_{opt}}}{NPSH3_{opt}^{0.75}}$$ where Q_{opt} is the flow rate per impeller eye at the best efficiency point, in m^3/s, and $NPSH3_{opt}$ is the corresponding value of NPSH3 for the first stage at maximum impeller diameter, in metres	$L^{0.75}T^{-1.5}$	–*
p	Pressure. In this book all pressures are gauge pressures, i.e. measured with respect to the atmospheric pressure, except the atmosperic pressure p_{amb}, the vapour pressure p_v and pressures referenced by p_{abs} which are absolute	$ML^{-1}T^{-2}$	Pa or bar†
p_{abs}	Absolute pressure; Sections 2.1 and 2.2 refer to the absolute pressure, designated by p_{abs}	$ML^{-1}T^{-2}$	Pa or bar†
p_{amb}	Atmospheric pressure	$ML^{-1}T^{-2}$	Pa or bar†
p_v	Vapour pressure of the fluid	$ML^{-1}T^{-2}$	Pa or bar†
Q	Volume rate of flow	L^3T^{-1}	m^3/s
ρ	Density, i.e. mass per unit volume	ML^{-3}	kg/m^3
SSS	Suction specific speed of the pump, in US units $$SSS = n\frac{\sqrt{Q_{opt}}}{NPSH3_{opt}^{0.75}}$$ where Q_{opt} is the flow rate per impeller eye at the best efficiency point, in USGPM, and $NPSH3_{opt}$ is the corresponding value of NPSH3 for the first stage at maximum impeller diameter, in feet. Conversion: $SSS = 51.64\ n_{ss}$	$L^{0.75}T^{-1.5}$	–*
u	Peripheral velocity in the impeller	LT^{-1}	m/s
v	Velocity of the fluid	LT^{-1}	m/s
z	Height above reference plane	L	m

*The numerical value of this parameter depends of the choice of the units.

†Refer to Appendix E for conversion factors: **bar** is the preferred unit in the pump industry.

Subscripts

Subscript	Designation	Example	
amb	Ambient	p_{amb}	Atmospheric pressure
1	At inlet of the pump	p_1	Pressure at pump inlet
2	At outlet of the pump	p_2	Pressure at pump outlet
v	Vapour	p_v	Vapour pressure
m	Meridional	c_m	Meridional component of absolute velocity
opt	At the best efficiency point (BEP)	Q_{opt}	Flow rate at BEP
min	Minimum	$p_{abs,min}$	Absolute pressure corresponding to the location in the pump with the lowest (minimum) pressure

Appendix A. Characteristics of water

T	p_v	ρ	c_p	μ	ν
°C	bar	kg/m^3	kJ/kg K	10^{-6} kg/m s	10^{-6} m^2/s
0.01	0.006112	999.8	4.217	1750	1.75
10	0.012271	999.7	4.193	1300	1.30
20	0.023368	998.3	4.182	1000	1.00
30	0.042417	995.7	4.179	797	0.800
40	0.073749	992.3	4.179	651	0.656
50	0.12334	988.0	4.181	544	0.551
60	0.19919	983.2	4.185	463	0.471
70	0.31161	977.7	4.190	400	0.409
80	0.47359	971.6	4.197	351	0.361
90	0.70108	965.2	4.205	311	0.322
100	1.0132	958.1	4.216	279	0.291
110	1.4326	950.7	4.229	252	0.265
120	1.9854	942.9	4.245	230	0.244
130	2.7012	934.6	4.263	211	0.226
140	3.6136	925.8	4.258	195	0.211
150	4.7597	916.8	4.310	181	0.197
160	6.1804	907.3	4.339	169	0.186
170	7.9202	897.3	4.371	159	0.177
180	10.003	886.9	4.408	149	0.168
190	12.552	876.0	4.449	141	0.161
200	15.551	864.7	4.497	134	0.155
210	19.080	852.8	4.551	127	0.149
220	23.201	840.3	4.614	122	0.145
230	27.979	827.3	4.686	116	0.140
240	33.480	813.6	4.770	111	0.136

T °C	p_v bar	ρ kg/m^3	c_p kJ/kg K	μ 10^{-6} kg/m s	ν 10^{-6} m^2/s
250	39.776	799.2	4.869	107	0.134
260	46.940	783.9	4.986	103	0.131
270	55.051	767.8	5.126	99.4	0.129
280	64.191	750.5	5.296	96.1	0.128
290	74.448	732.1	5.507	93.0	0.127
300	85.917	712.2	5.773	90.1	0.127
310	98.697	690.6.	6.120	86.5	0.125
320	112.90	666.9	6.586	83.0	0.124
330	128.65	640.5	7.248	79.4	0.124
340	146.08	610.3	8.270	75.4	0.124
350	165.37	574.5	10.08	70.9	0.123
360	186.74	528.3	14.99	65.3	0.124
370	210.53	448.3	53.92	56.0	0.125
374.15	221.20	315.5		45.0	0.143

T, Celsius temperature; c_p, specific heat at constant pressure; p_v, vapour pressure; μ, dynamic viscosity; ρ, density; ν, kinematic viscosity.

Appendix B. NPSH reduction for pumps handling hydrocarbon liquids and high temperature water

Explanation for the use of Figure B1

Figure B1 is a composite chart of NPSH reductions which may be expected for hydrocarbon liquids and high temperature water, based on available laboratory data from tests conducted on the fluids shown, plotted as a function of fluid temperature and vapour pressure at that temperature.

The following limitations and precautions should be observed in the use of Figure B1.

Until specific experience has been gained with operation of pumps under conditions where this chart applies. NPSH reduction should be limited to 50% of the NPSH required by the pump for cold water.

This chart is based on pumps handling pure liquids. Where entrained air or other non-condensable gases are present in a liquid, pump performance may be adversely affected even with normal NPSH available and would suffer further with reduction in NPSH available. Where dissolved air or other noncondensables are present and where the absolute pressure at the pump inlet would be low enough to release such

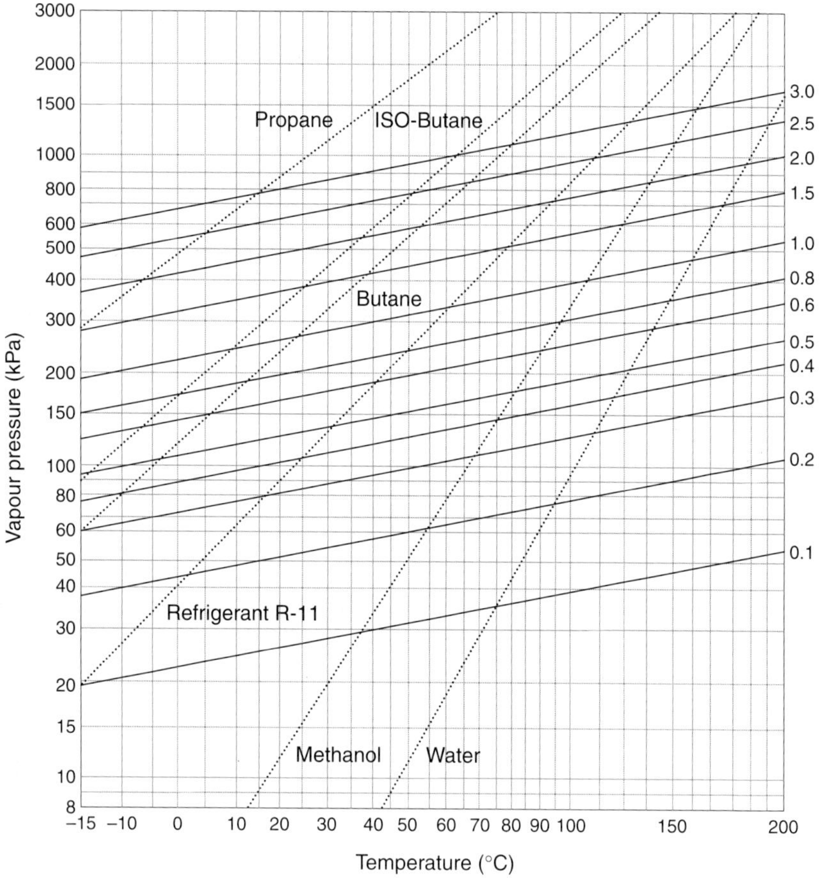

Figure B1 NPSH reduction for pumps handling hydrocarbon liquids and high temperature water (the table is based on Hydraulic Institute Standards [HIS]. 1985).

Note 1: Available data are limited to the liquids for which the temperature–vapour pressure relationship is shown on the figure. Application of this chart to liquids other than hydrocarbons and water is not recommended without experimental basis.

Note 2: The NPSH reduction actually applied to the NPSH required in cold water shall be either the value read on the right scale of the chart, or one half the NPSHR in cold water, whichever is the smaller.

noncondensables from solution, the NPSH available may have to be increased above that required for cold water to avoid deterioration of pump performance due to such release.

For hydrocarbon mixtures, vapour pressure may vary significantly with temperature and specific vapour pressure determinations should be made for actual pumping temperatures.

In the use of the chart for high temperature liquids, and particularly with water, due consideration must be given to the susceptibility of the suction system to transient changes in temperature and absolute pressure, which might necessitate provision of a margin of safety of NPSH far exceeding the reduction otherwise available for steady state operation.

Because the absence of available data demonstrating NPSH reduction greater than 3 m, the chart has been limited to that extent and extrapolation beyond that limit is not recommended.

Enter Figure B1 at the bottom of the chart with pumping temperature in °C and proceed vertically upward to the vapour pressure. From this point follow along or parallel to the sloping lines to the right side of the chart, where the NPSH reduction may be read on the scale provided. If this value is greater than one half of the NPSH required on cold water, deduct one half of the cold water NPSH to obtain corrected NPSH required. If the value on the chart is less than one half of the cold water NPSH, deduct this chart value from the cold water NPSH to obtain corrected NPSH required.

Appendix C. Abstract from ISO/ CD 9906

Type of installation	Open sump	Open sump	Open sump
Independent variable	Inlet throttle valve	Outlet throttle valve	Water level
Constant	Outlet throttle valve	Inlet throttle valve	Inlet and Outlet throttle valves
Quantities the variation of which is dependent on control	Total head, rate of flow, NPSHA, water level	Total head, rate of flow, NPSHA, water level	Total head, rate of flow NPSHA

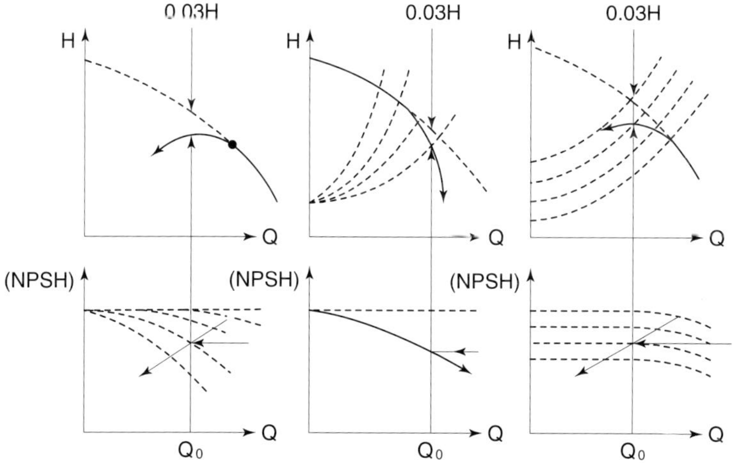

Type of installation	Open sump	Open sump	Closed loop	Closed loop
Independent variable	Inlet throttle valve	Water level	Pressure in the tank	Temperatuure (vapour pressure)
Constant	Flow rate	Flow rate	Flow rate	Flow rate
Quantities the variation of which is dependent on control	NPSHA, total head, outlet, throttle valve (for constant rate of flow)	NPSHA, total head, outlet throttle valve	Total head, NPSHA, outlet throttle, valve (for constant rate of flow, when total head begins to drop)	NPSHA, outlet throttle valve (for constant rate of flow, when head begins to drop)

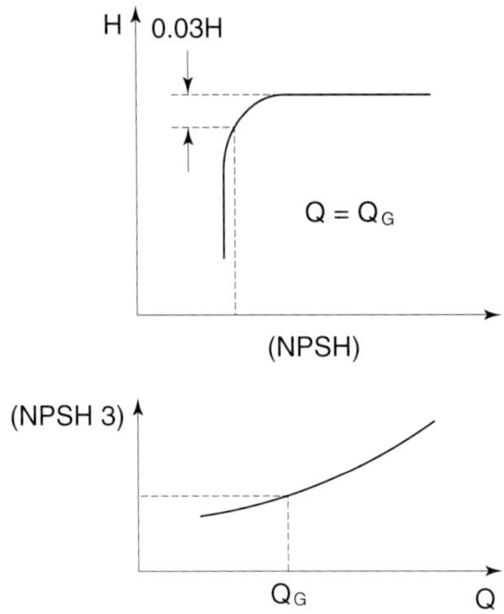

Type of installation	Closed loop	Closed sump or loop
Independent variable	Pressure in the tank	Temperature (vapour pressure)
Constant	Inlet and Outlet throttle valves	
Quantities the variation of which is dependent on control	● NSPHA ● total head and rate of flow, when a certain level of cavitation is reached.	

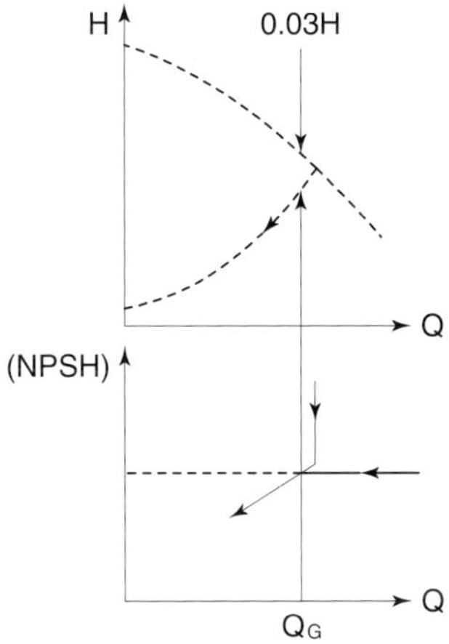

Appendix D. Definition of the total suction head H_A (abstract from ISO 2548)

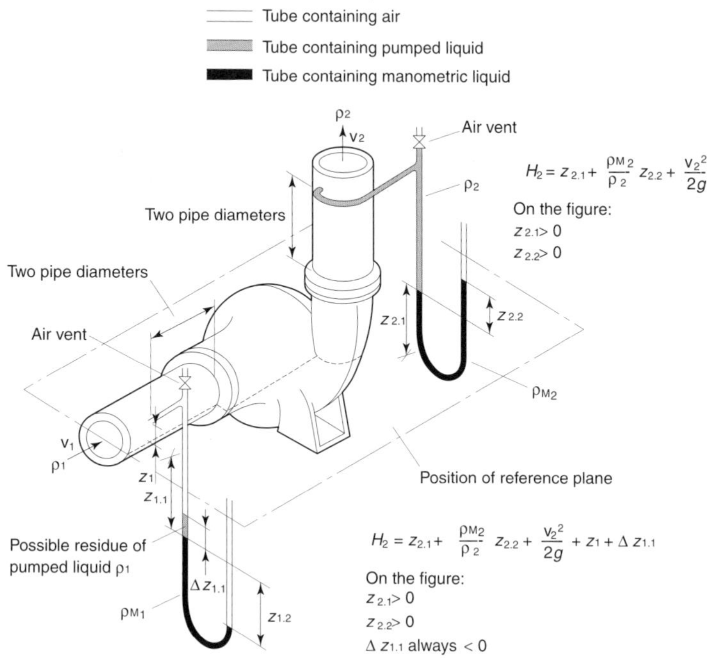

(a) The pump inlet is under vacuum

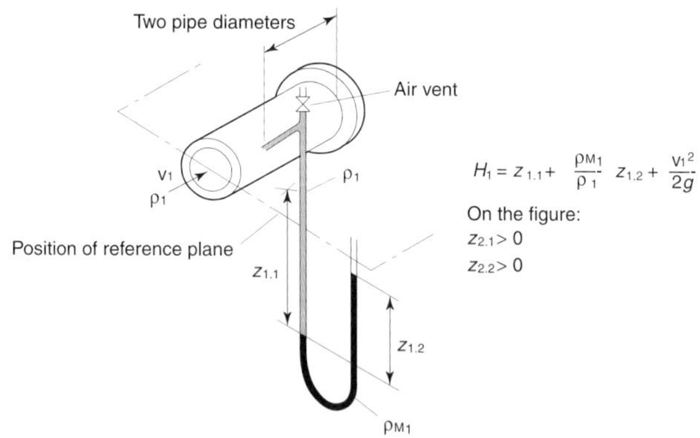

(a) The pump inlet is under pressure

Figure D1 Test of a centrifugal pump by means of liquid column gauges. The drawings show the principle but not the full technical details.

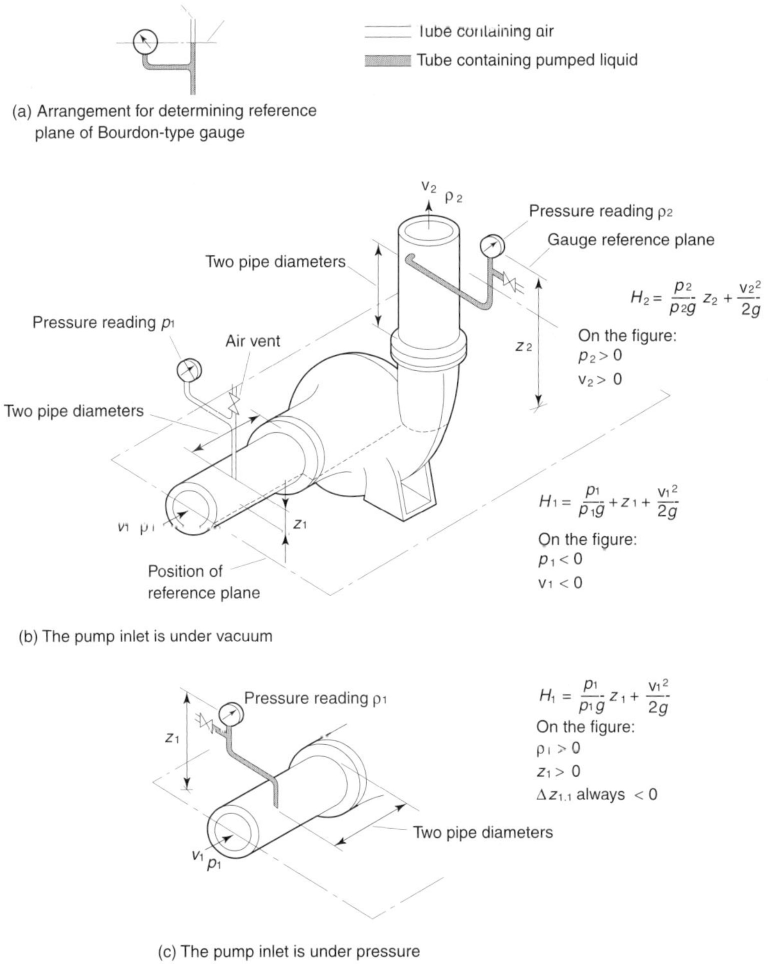

Figure D2 Test of a centrifugal pump by means of Bourdon gauges. The drawings show the principle but not the full technical details.

$$H_1 = z_1$$

$$H_2 + \frac{p_2}{\rho_2 g} + z_2 + \frac{v_2^2}{2g}$$

$$H_2 + \frac{p_2}{\rho_2 g} + z_{1\cdot 2} + \frac{v_2^2}{2g}$$

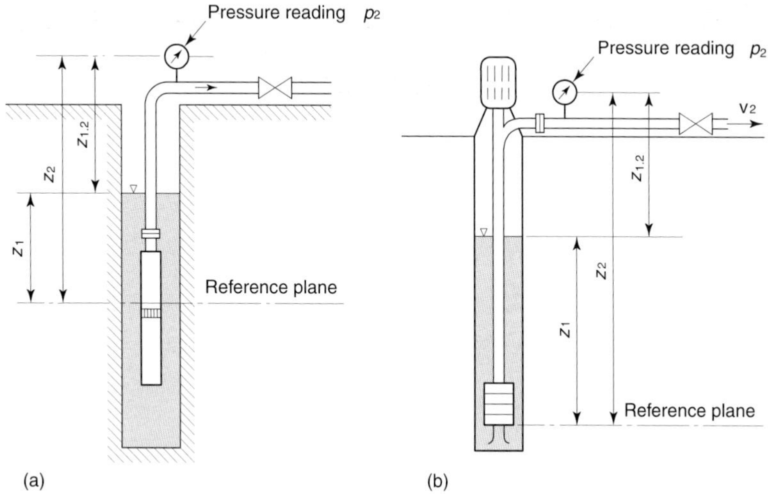

Figure D3 Measurement of pump total head H for various types of submerged pumps.

Appendix E. Table of conversion for pressure

	N/m^2 = Pa	bar	μbar	kp/m^2	at	atm	Torr	psi
N/m^2 = 1 Pa	1	$10^{\times 5}$	10	1.019716×10^{-1}	1.019716×10^{-5}	0.986923×10^{-5}	0.750062×10^{-2}	1.4503×10^{-4}
bar = 106dyn/cm^2	10^5	1	10^6	1.019716×10^4	1.01976	0.986923	0.750062×10^3	1.4503×10^1
μbar = 1 dyn/cm^2	10^{-1}	10^{-6}	1	1.019716×10^{-2}	1.019716×10^{-6}	0.986923×10^{-6}	0.750062×10^3	1.4503×10^{-5}
kp/m^2	0.980665×10	0.980665×10^{-4}	0.980665×10^{-2}	1	10^{-4}	0.967841×10^{-4}	$0.735559 \times 10-1$	$1.4223 \times 10-3$
at = 1 kp/cm^2 =	0.980665×10^5	0.980665	0.980665×10^6	10^4	1	0.967841	0.735559×10^3	14.223
atm = 760 Torr =	1.01325×10^5	1.01325	1.01325×10^6	1.033227×10^4	1.033227	1	0.760×10^3	14.695
Torr	1.333224×10^2	1.333224×10^{-3}	1.333224×10^3	13.59510	1.359510×10^{-3}	1.315789×10^{-3}	1	1.9336×10^{-2}
1 psi = pound per square inch	6.895×10^3	0.06895	6.895×10^4	7.031×10^2	0.7031	0.06805	0.51718×10^2	1